Numbers

© 2015 OnBoard Academics, Inc
Portsmouth, NH
800-596-3175
www.onboardacademics.com
ISBN: 978-1-63096-065-0

OnBoard Academic's books are specifically designed to be used as printed workbooks or as on-screen instruction. Each page offers focused exercises and students quickly master topics with enough proficiency to move on to the next level.

OnBoard Academic's lessons are used in over 25,000 classrooms to rave reviews. Our lessons are aligned to the most recent governmental standards and are updated from time to time as standards change. Correlation documents are located on our website. Our lessons are created, edited and evaluated by educators to ensure top quality and real life success.

Interactive lessons for digital whiteboards, mobile devices, and PCs are available at www.onboardacademics.com. These interactive lessons make great additions to our books.

You can always reach us at customerservice@onboardacademics.com.

Skip Counting

Key Vocabulary

Skip Counting

Multiply

Multiple

Skip Counting by Twos.
Write the number of boxes in the space provided and then count by twos.

Skip Count by 2s

Skip Counting by Fives.

Write the number of boxes in the space provided and then count by fives.

	Skip Count by 5s

Skip Counting by Tens

Write the number of boxes in the space provided and then count by tens.

Practice by writing in the missing values.

28		32	34		38		42

20		30	35			50	55

20		40	50			80	

Write the red numbers in the correct positions.

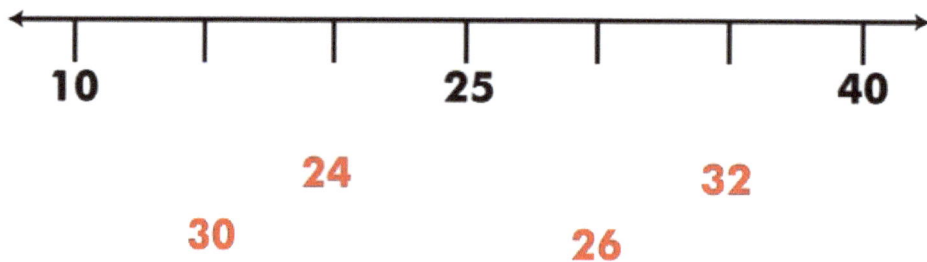

15   20   30   35

22   28   34

10   25   40

24   32

30   26

Use the 100 square to practice skip counting. Circle the numbers as you count.

Skip count by 2, 3, 5, and 10

1	2	3	4	5	6	7	8	9	10
11	12	13	14	15	16	17	18	19	20
21	22	23	24	25	26	27	28	29	30
31	32	33	34	35	36	37	38	39	40
41	42	43	44	45	46	47	48	49	50
51	52	53	54	55	56	57	58	59	60
61	62	63	64	65	66	67	68	69	70
71	72	73	74	75	76	77	78	79	80
81	82	83	84	85	86	87	88	89	90
91	92	93	94	95	96	97	98	99	100

1	2	3	4	5	6	7	8	9	10
11	12	13	14	15	16	17	18	19	20
21	22	23	24	25	26	27	28	29	30
31	32	33	34	35	36	37	38	39	40
41	42	43	44	45	46	47	48	49	50
51	52	53	54	55	56	57	58	59	60
61	62	63	64	65	66	67	68	69	70
71	72	73	74	75	76	77	78	79	80
81	82	83	84	85	86	87	88	89	90
91	92	93	94	95	96	97	98	99	100

1	2	3	4	5	6	7	8	9	10
11	12	13	14	15	16	17	18	19	20
21	22	23	24	25	26	27	28	29	30
31	32	33	34	35	36	37	38	39	40
41	42	43	44	45	46	47	48	49	50
51	52	53	54	55	56	57	58	59	60
61	62	63	64	65	66	67	68	69	70
71	72	73	74	75	76	77	78	79	80
81	82	83	84	85	86	87	88	89	90
91	92	93	94	95	96	97	98	99	100

1	2	3	4	5	6	7	8	9	10
11	12	13	14	15	16	17	18	19	20
21	22	23	24	25	26	27	28	29	30
31	32	33	34	35	36	37	38	39	40
41	42	43	44	45	46	47	48	49	50
51	52	53	54	55	56	57	58	59	60
61	62	63	64	65	66	67	68	69	70
71	72	73	74	75	76	77	78	79	80
81	82	83	84	85	86	87	88	89	90
91	92	93	94	95	96	97	98	99	100

Name:_____

Skip Counting Quiz

Circle or fill in the correct answer.

1 **True or false, the missing number in this skip counting sequence is 30?**   5, 10, 15, 20, 25, ___, 35

2 **Which skip counting sequence is not correct?**

A 12, 14, 16, 18, 20

B 35, 40, 50, 60, 70

C 2, 4, 6, 8, 10, 12

D 20, 30, 40, 50, 60

3 2 x ___ = 16

4 5 x ___ = 25

Compare and Order Numbers

Key Vocabulary

compare

order

less than

greater than

www.onboardacademics.com

Draw the red numbers onto the number line.

30 ——————————————————————— 50

(40)   (35)   (47)

_____

( )( ) Can you think of two more numbers that **would appear** on this number line, and two numbers **that wouldn't?** ( )( )

Write the numbers that would appear in the red circles and the numbers that wouldn't appear on the number line in the blue circles.

Comparing Numbers

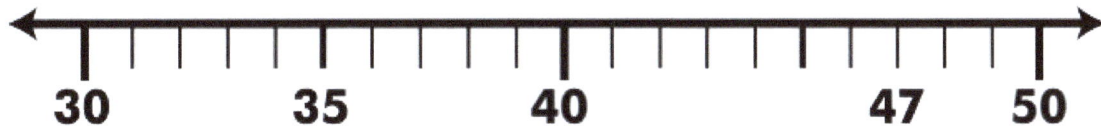

1	Which number is smaller, 40 or 47?	
2	Which number is larger, 40 or 47?	
3	Which number is smaller, 35 or 50?	
4	Which number is larger, 35 or 50?	

www.onboardacademics.com

Locate these numbers on the number line. Draw them in their proper place.

100   105   92

90   110

_____

Is 92 smaller or larger than 105?

Write your answer here_____

_____

Can you think of two more numbers that **would appear** on this number line, and two numbers **that wouldn't?**

Write your answers in the red and blue boxes.

Locate 400 on the number line. Draw it in place. Now sort the numbers by writing into one of the boxes.

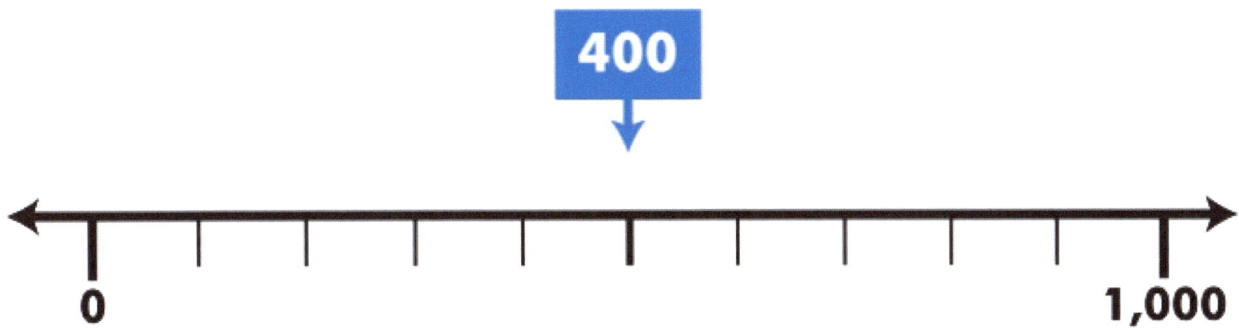

400

0   1,000

900 300 800 200 700 500 100 600

Less than 400	Greater than 400

Less than, greater than, equal to

Study the illustration below. identify the numbers on the number line, reach the comparative statement and then look at the corresponding equation.

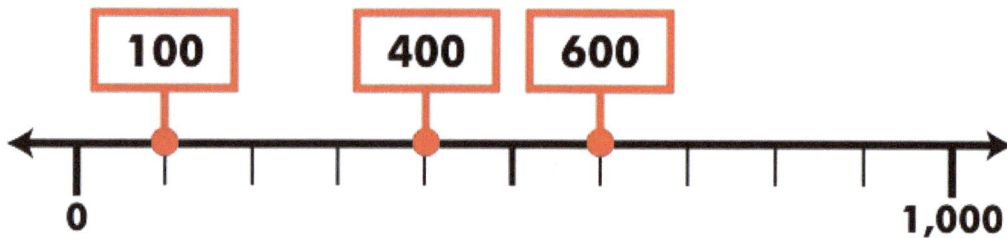

| 100 | 400 | 600 |

0   1,000

100 is less than 400   **100 < 400**

600 is greater than 400   **600 > 400**

1,000 is equal to 1,000   **1,000 = 1,000**

_____

How can I remember this?

27   <   54

54   >   27

Look at the next page for a hint.

_____

less than

27 < 54

greater than

54 > 27

"I always eat the biggest number"

_____

Less than, greater than and equal to practice.
Use the symbols below to complete the equation.

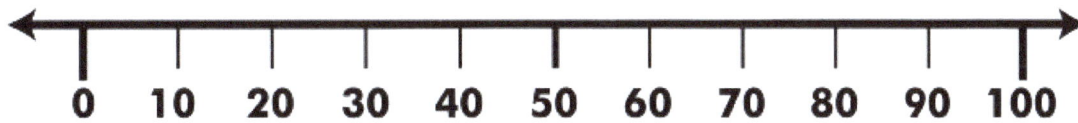

0   10   20   30   40   50   60   70   80   90   100

| 54 [ ] 63 | 84 [ ] 73 |

| 83 [ ] 100 | 89 [ ] 98 |

| 25 [ ] 25 | 17 [ ] 16 |

< = >

More practice, fill in the blanks.

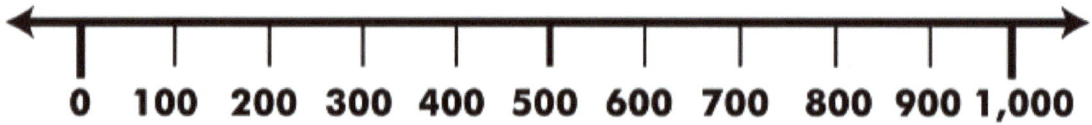

154		163
883		104
784		372

365		365
889		898
101		99

| < | = | > |

Name_____

Compare and Order Numbers Quiz

Circle the correct answer.

1 $17 < 19$ **True or false.**

2   **Which statement is not correct?**

     A $34 > 21$

     B $87 < 78$

     C $123 > 103$

     D $844 < 944$

3   **Which number is the smallest?**   74   64   91   94

4   **Which number is the largest?**   145   514   414   235

Compare and Order Numbers

Key Vocabulary

Less than

Greater than

Equal to

Compare

Order

Who scored the most points this season?

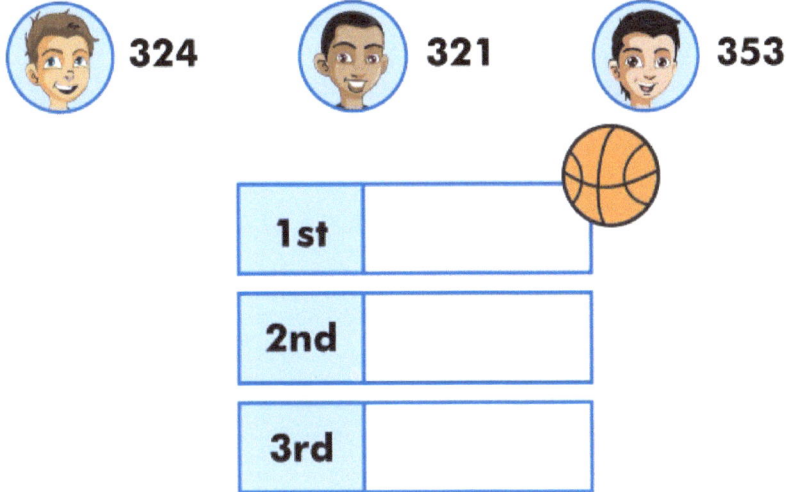

324   321   353

1st	
2nd	
3rd	

1. Compare hundreds — they're the same

2. Compare tens — which is the largest number?

3. Compare ones — which is the largest number?

hundreds	tens	ones
3	2	1
3	5	3
3	2	4

Order these numbers from least to greatest.

thousands	hundreds	tens	ones
4,	3	2	7
1,	4	2	4
4,	9	8	4
1,	2	3	4

Least to greatest ⟶

Least   Greatest

Order these numbers on the number line.
Write the correct number in the box on the number line.

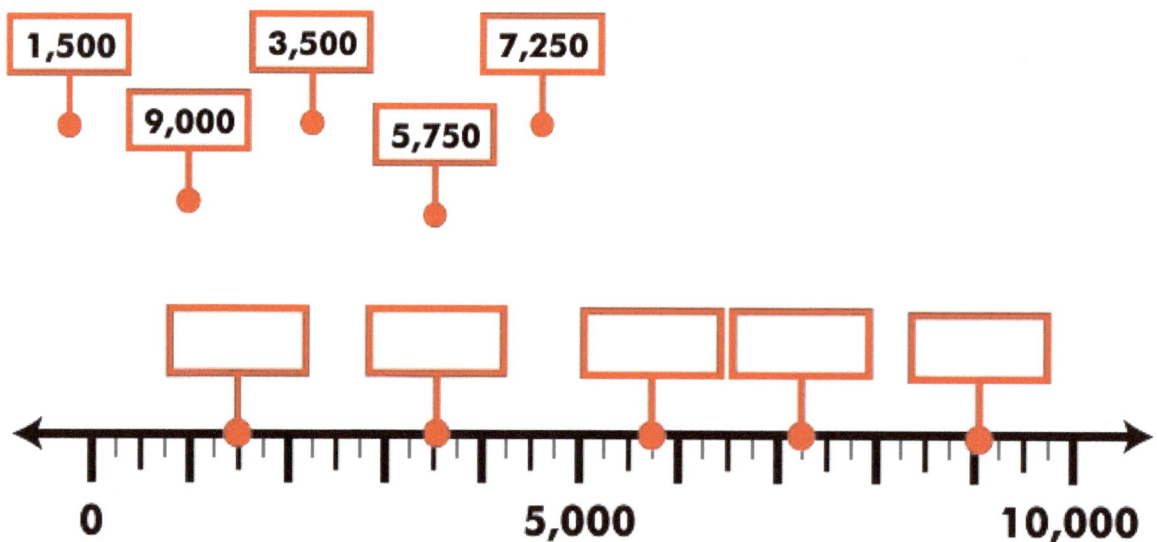

1,500   3,500   7,250

9,000   5,750

0   5,000   10,000

Less than, greater than and equal to.
Write the correct symbol in the box.

84 ☐ 48	900 ☐ 999
108 ☐ 184	1,349 ☐ 2,349
7 x 5 ☐ 35	9,999 ☐ 7,721

| < | = | > |

Name_____

Compare and Order Numbers Quiz

Circle the correct answer.

1 True or false? 5,432 > 5,379

2   **3**   **4**

Which of these statements is *not* correct?

A 354 > 259	A 654 > 564	A 9,876 > 8,796
B 793 > 973	B 456 < 654	B 7,968 > 7,896
C 459 > 399	C 564 < 465	C 9,786 > 9,867
D 9 x 8 = 72	D 645 > 546	D 7,986 > 7,689

Odd & Even Numbers

Key Vocabulary

Pairs

Odd numbers

Even numbers

Sort the odd and even cards. Draw the cards in the empty boxes.

GROUP 1	GROUP 2

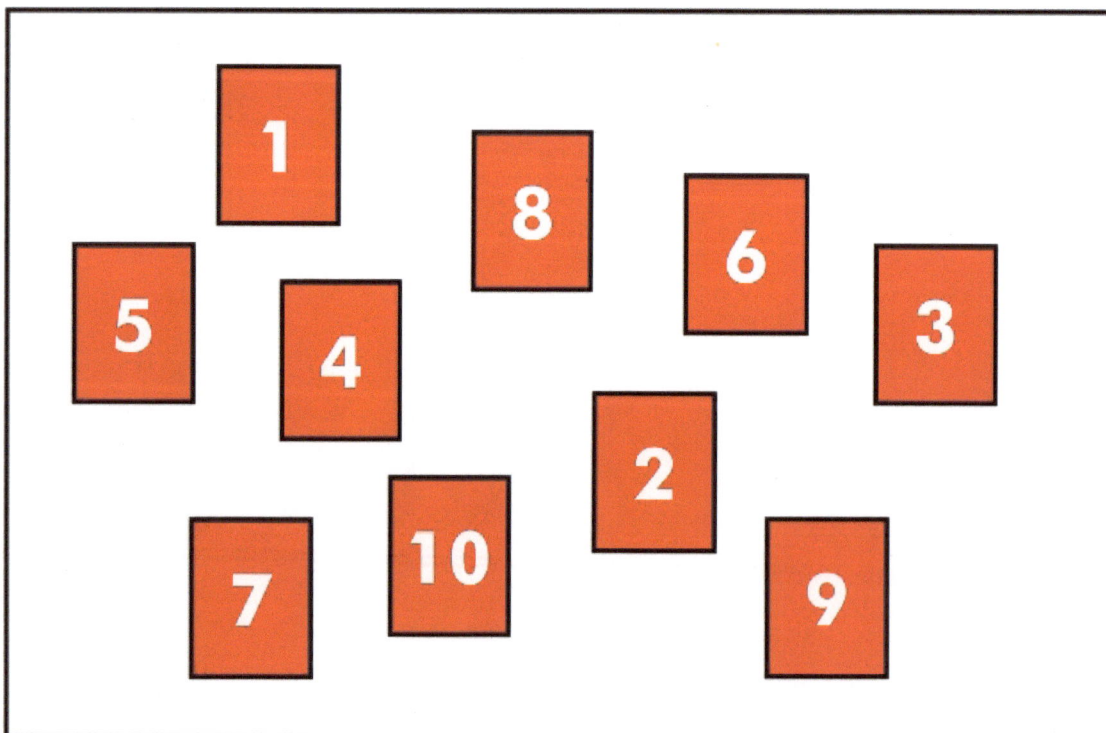

1   8   6   3

5   4

2

7   10   9

Partner up the students by connecting them with a line.
Any student who doesn't have a partner, connect them with the green box.

Draw a line between the number and the corresponding number icon.

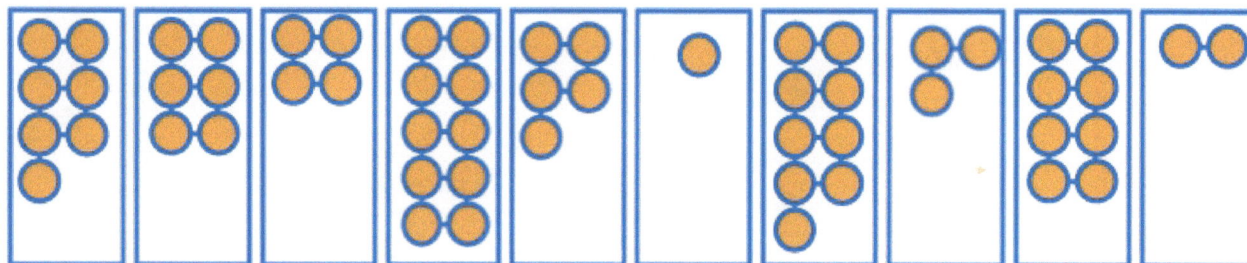

| 1 | 2 | 3 | 4 | 5 | 6 | 7 | 8 | 9 | 10 |

Why do you think these numbers are called "odd" numbers?

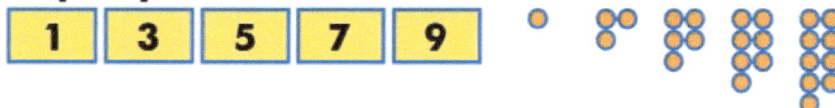

| 1 | 3 | 5 | 7 | 9 |

_____

_____

Why do you think these numbers are called "even" numbers?

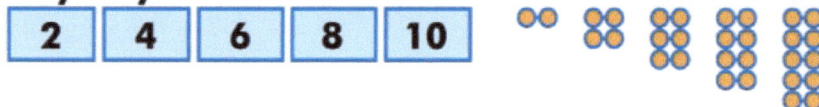

| 2 | 4 | 6 | 8 | 10 |

_____

_____

Find the missing numbers and write it in the space provided.

Odd or even? Write the number in the proper box.

Odd Numbers	1, 3, 5, 7,	☐	11, 13,	☐	17, 19,	☐

Even Numbers	2, 4, 6, 8,	☐	12, 14,	☐	18, 20,	☐

ODD NUMBERS	EVEN NUMBERS

52　29　32　　100　41　71

28　57　46　13　94　63　95　44

What do you notice? Fill in the blue and gold boxes.

ODD NUMBERS
95   63
57
71   13
29
41

EVEN NUMBERS
100   94
52
46
44   28
32

Odd numbers always end in ☐ ☐ ☐ ☐ ☐

Even numbers always end in ☐ ☐ ☐ ☐ ☐

Name_____

Odd & Even Quiz

Circle the correct answer.

1 True or false? There is an even number of boys and girls.

2 Even numbers always end with a digit of:

A 2, 4, 6, or 8

B 0, 2, 4, 6, or 8

C 1, 3, 5, 7, or 9

D 3, 5, 7, or 9

3 Which number is an odd number? 154 889 304 990

4 Which number is an even number? 71 89 65 90 111

Ordinal Numbers

Key Vocabulary

first

second

third

fourth

fifth

sixth

seventh

eighth

ninth

tenth

What is the current position of the cars on this road? If you have crayons draw the color in the correct box, if not draw a line or write the name of the color.

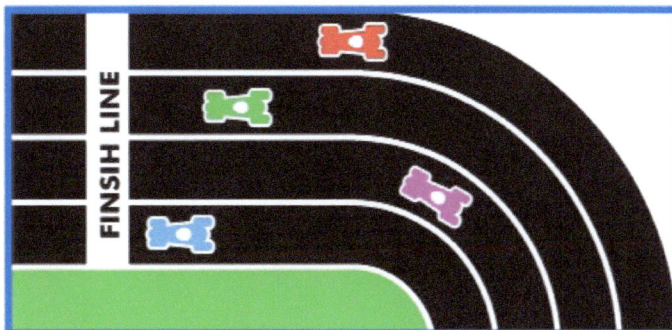

1st place	2nd place	3rd place	4th place

Complete this chart with the suggestions provided to see the first 10 ordinal numbers and their abbreviations. Study the chart.

Ordinal Number	Abbreviation
First	1st
Second	2nd
	3rd
Fourth	4th
	5th
	6th
	7th
Eighth	8th
	9th
Tenth	10th

Seventh

Ninth

Sixth

Third

Fifth

Draw a line from the ordinal number on the bottom to the correct date. The first one is completed for you.

By the way, do you know what is special about the date in yellow? _____

February 2008

S	M	T	W	T	F	S
					1	2
3	4	5	6	7	8	9 Dad's 50th
10	11 Dentists	12	13	14	15	16
17	18	19	20	21	22 Business trip to Madison, WI	23
24	25 Golf	26	27	28	29	

eighth	sixth	fifth	tenth	seventh

Ordinal number word scramble.
Unscramble the ordinal numbers and then write in the abbreviation. The first one is done for you as an example.

DIRTH

THIRD	3rd

HEIGHT

HINNT

THORFU

HETTN

STRIF

Order the students by writing in their place.
Mia came in first in the math test. What place did the other students come in?

James	88	
Owen	84	
Mia	97	⭐ **First**
Alison	79	
Tony	93	

Name_____

Ordinal Number Quiz

Circle or fill in the correct answer.

1 True or false? The fifth apple has been bitten.

2 Which apple has been bitten?

- **A** First
- **B** Eighth
- **C** Tenth
- **D** Ninth

3 What was the test score of the person who finished third?

4 What was the test score of the person who finished fifth?

| Test Scores | 47 | 33 | 49 | 27 | 48 | 44 |

Place Value

Key Vocabulary

Digit

Ones

Tens

Hundreds

What value is represented by each of these blocks? Write the answer in the box.

What number do the blocks represent? Hint; count the number of rows of ten blocks and then the number of single blocks (green).

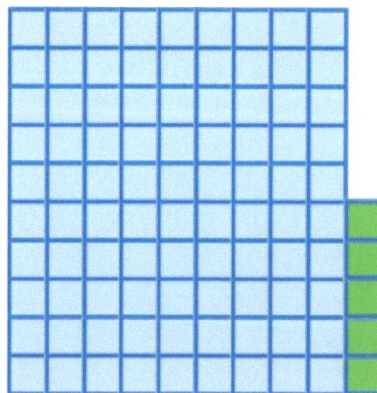

tens	ones	Answer

+ =

What number do these blocks represent?

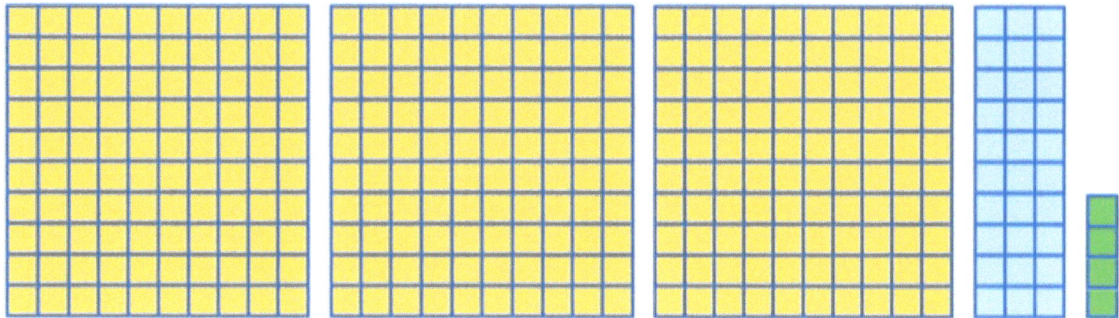

hundreds + tens + ones = Answer

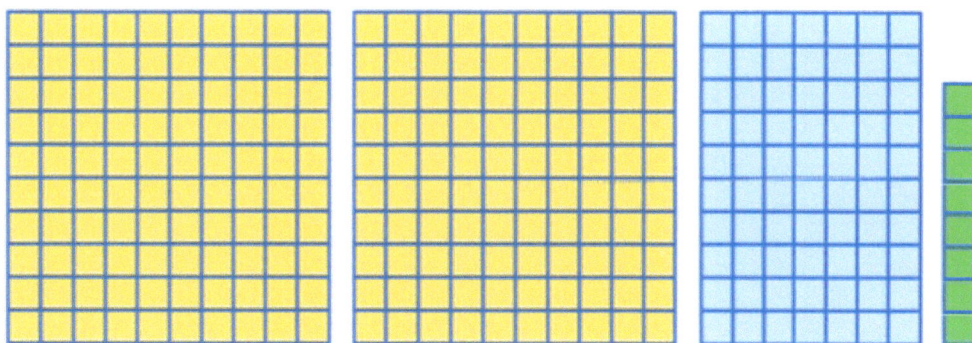

hundreds + tens + ones = Answer

OnBoard Academics Workbook   K-2 Mathematics

How many of each of the 100 blocks, ten blocks and ones blocks do you need to make these numbers? Write the answer in the space provided.

	hundreds +		tens +		ones =	**230**

	hundreds +		tens +		ones =	**333**

What is the value of the red 5? The first one is done for you.

hundreds tens ones

3 5 7   **2 7 5**   **5 7 0**

| 50 | | |

Name_____

Place Value Quiz

Circle or fill in the correct answer.

1   **8 6 5** = 80 + 60 + 5

2   **What number do the blocks in Figure 1 represent?**

Ⓐ 66

Ⓑ 68

Ⓒ 58

Ⓓ 108

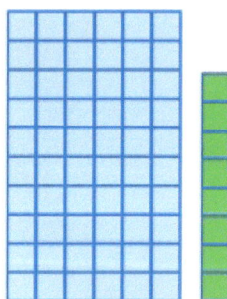

Figure 1

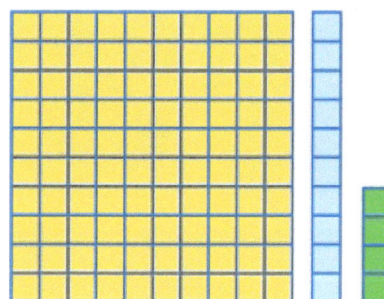

Figure 2

3   **What number do the blocks in Figure 2 represent?**

4   714 = 700 + ____ + 4

Place Value

Key Vocabulary

Less than

Greater than

Equal to

Compare

Order

Which Island is the hottest?
Draw the sun over the hottest island.

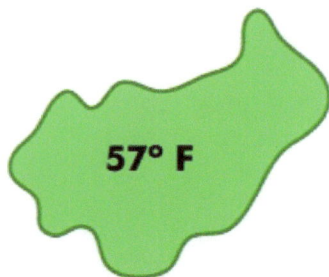

Santa de Sizzla

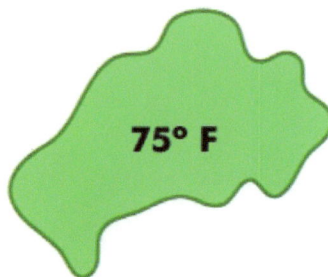

Isle de Scorchio

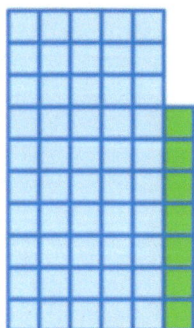

5 tens + 7 ones

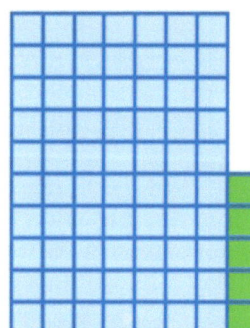

7 tens + 5 ones

Which product is the Star Buy?

The Star Buy is the best or lowest price. Draw a star in the product's circle to indicate the Star Buy.

$463	⭐ STAR BUY!	$436
$1,505	⭐ STAR BUY!	$1,550

Investigate place value.

Fill in the boxes below

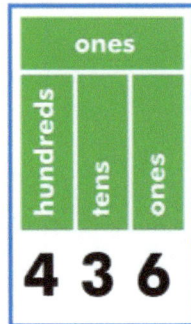

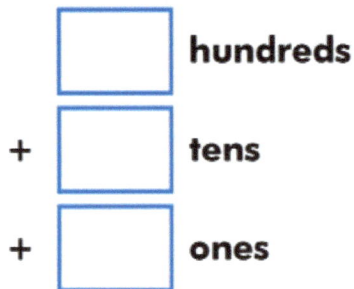

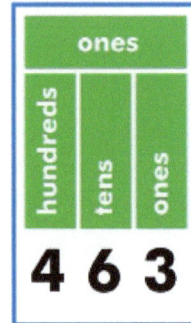

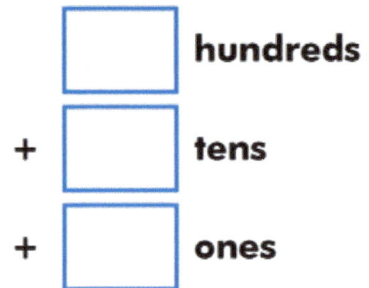

	ones	
hundreds	**tens**	**ones**
4	3	6

☐ hundreds

+ ☐ tens

+ ☐ ones

	ones	
hundreds	**tens**	**ones**
4	6	3

☐ hundreds

+ ☐ tens

+ ☐ ones

Use the symbols < > to compare the weight of Ella and Eli.

Write the weight of the smallest elephant in the box above to identify ones, tens, hundreds and thousands.

Ella	Eli
4,5 4 2 lb	4,4 5 2 lb

thousands			ones		
hundreds	tens	ones	hundreds	tens	ones

< less than

> greater than

_____

Investigate place value with thousands.

1,5 0 5   1,5 5 0

[ ]	[ ]	thousands
+ [ ]	+ [ ]	hundreds
+ [ ]	+ [ ]	tens
+ [ ]	+ [ ]	ones

thousands			ones		
hundreds	tens	ones	hundreds	tens	ones

Do you get < and > mixed up.
Use these alligators giant jaws as a reminder. The alligator always eats the bigger number.

greater than

452 > 341

341 < 452

less than

"I always eat the biggest number"

What are the largest numbers and the smallest numbers that you can make with the digits 3, 5, and 9.

ones		
hundreds	tens	ones
			largest
			smallest

3 5 9

Fill in the correct symbol.

| 1 | 689 | ○ | 986 | | 2 | 24,834 | ○ | 32,834 |

| 3 | 1,324 | ○ | 1,234 | | 4 | 65,234 | ○ | 65,434 |

| 5 | 11,324 | ○ | 11,624 | | 6 | 74,234 | ○ | 77,004 |

(<)   (>)

| 1 | 45 | ○ | 54 | | 2 | 89 | ○ | 78 |

| 3 | 101 | ○ | 110 | | 4 | 735 | ○ | 753 |

| 5 | 463 | ○ | 436 | | 6 | 989 | ○ | 898 |

(<)   (>)

Name_____

Place Value Quiz

Circle or fill in the correct answer.

1 True or false? $945 < $954

2 Write 9 hundreds + 5 tens + 0 ones in standard form.
- **A** 905
- **B** 950
- **C** 509
- **D** 590

3 Write 5 thousands + 0 hundreds + 0 tens + 3 ones in standard form.

4 Write 9 thousands + 9 hundreds + 4 tens + 7 ones in standard form.